FUSION

FARMER LLAMA'S FARM MACHINES

COMBINES

BY KIRSTY HOLMES

BEARPORT
PUBLISHING

Minneapolis, Minnesota

Library of Congress Cataloging-in-Publication Data is available at www.loc.gov or upon request from the publisher.

ISBN: 978-1-64747-543-7 (hardcover)
ISBN: 978-1-64747-550-5 (paperback)
ISBN: 978-1-64747-557-4 (ebook)

© 2021 Booklife Publishing
This edition is published by arrangement with Booklife Publishing.

For more information, write to Bearport Publishing, 5357 Penn Avenue South, Minneapolis, MN 55419. Printed in the United States of America.

IMAGE CREDITS

All images are courtesy of Shutterstock.com, unless otherwise specified. With thanks to Getty Images, Thinkstock Photo, and iStockphoto.
Cover - NotionPic, Tartila, A-R-T, logika600, BiterBig, Hennadii. Aggie - NotionPic, Tartila. Grid - BiterBig. Farm - Faber14. Spreaders - Hennadii H. 2 - alazur.
4 - Mascha Tace. 6-7 - Hennadii H, K-Nick. 6 - YRABOTA. 7 - ANEK SANGKAMANEE, Judith Lienert. 8 - Nsit, firtinali. 9 - Magicleaf. 10 - judyjump, Hennadii H.
12 - Katason. 13 - panotthorn phuhual, ananaline, NotionPic, naimtastik. 14 - studioworkstock. 15 - Hennadii H, Varga Jozsef Zoltan. 16 - Scharfsinn.
17 - Rachael Arnott. 19 - BigMouse. 21 - ArtMalivanov, DRogatnev. 22 - AVIcon, alexandrovskyi, Olha Bocharova, garykingphotographer. 23 - shaineast.

CONTENTS

Down on the Farm! 4

What Is a Combine? 6

Before Combines 8

Let's Combine All Three! 12

Parts of a Combine 14

Combines of the Future............... 16

Record Breakers...................... 18

Get Your Llama-Diploma 20

Going in Circles 22

Glossary 24

Index................................. 24

DOWN ON THE FARM!

You must be the new **farmhand**! Thank goodness you've arrived. Harvest time is a busy time here at Happy Valley Farm.

My name is Aggie. I'm a farmer llama. I'll be teaching you all about combine harvesters, or combines, today!

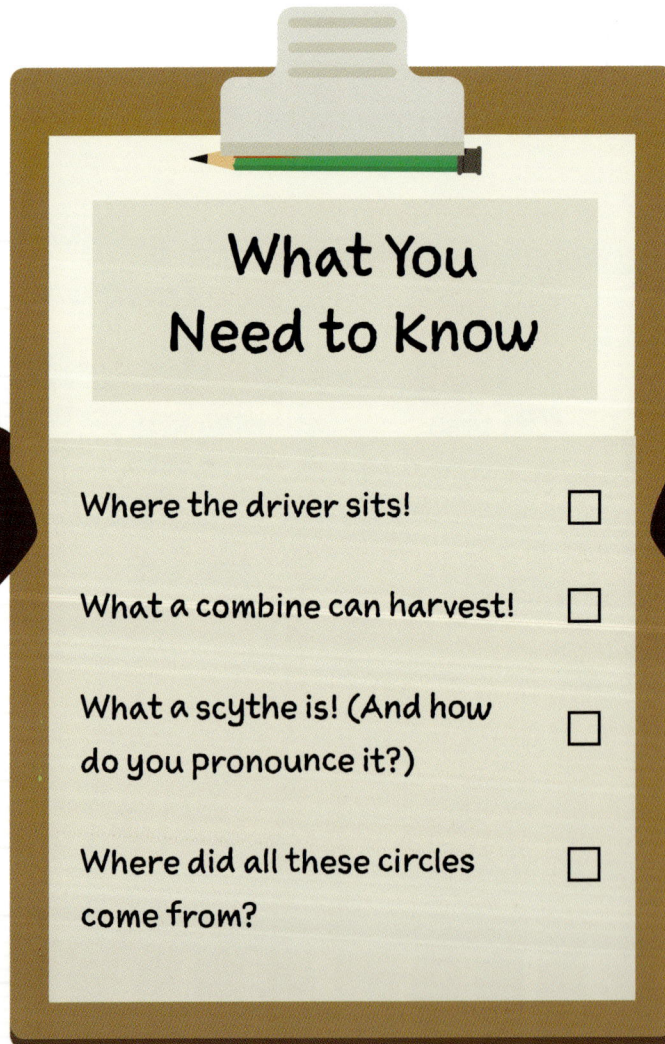

What You Need to Know

Where the driver sits! ☐

What a combine can harvest! ☐

What a scythe is! (And how do you pronounce it?) ☐

Where did all these circles come from? ☐

WHAT IS A COMBINE?

A combine is a machine that is found on a farm. It is very important for any farmer who grows **crops**. Combines are used to gather crops from fields.

WHEAT

Some of the world's most important foods are crops. Corn, rice, wheat, potatoes, soybeans, and many other crops are grown across the world.

CORN

RICE

Many of our favorite foods are gathered using combines.

BEFORE COMBINES

Before combines, all farmwork was done by humans and animals. When a crop was fully grown and ready to eat, there were three main things that needed to be done.

REAPING

Cutting the crop down is called reaping. This can be done with a sickle or a scythe.

SICKLE

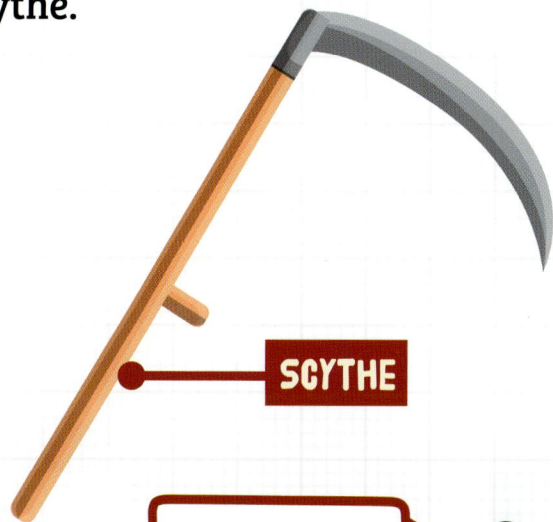

SCYTHE

Say it siTHe!

THRESHING

The grain is the bit you can eat. Farmers pull it apart from the stalk. This is called threshing.

9

WINNOWING

Winnowing throws the grain into the air so that any dry parts can blow away. Heavy grains stay behind to be used. Winnowing is also known as cleaning.

During the harvest, children took time off school.

People harvested crops like this for a long time.
Now, machines can be used to speed things up.

11

LET'S COMBINE ALL THREE!

Combines can do all of the main harvesting jobs: reaping, threshing, and winnowing. This is much faster than doing the jobs **by hand**.

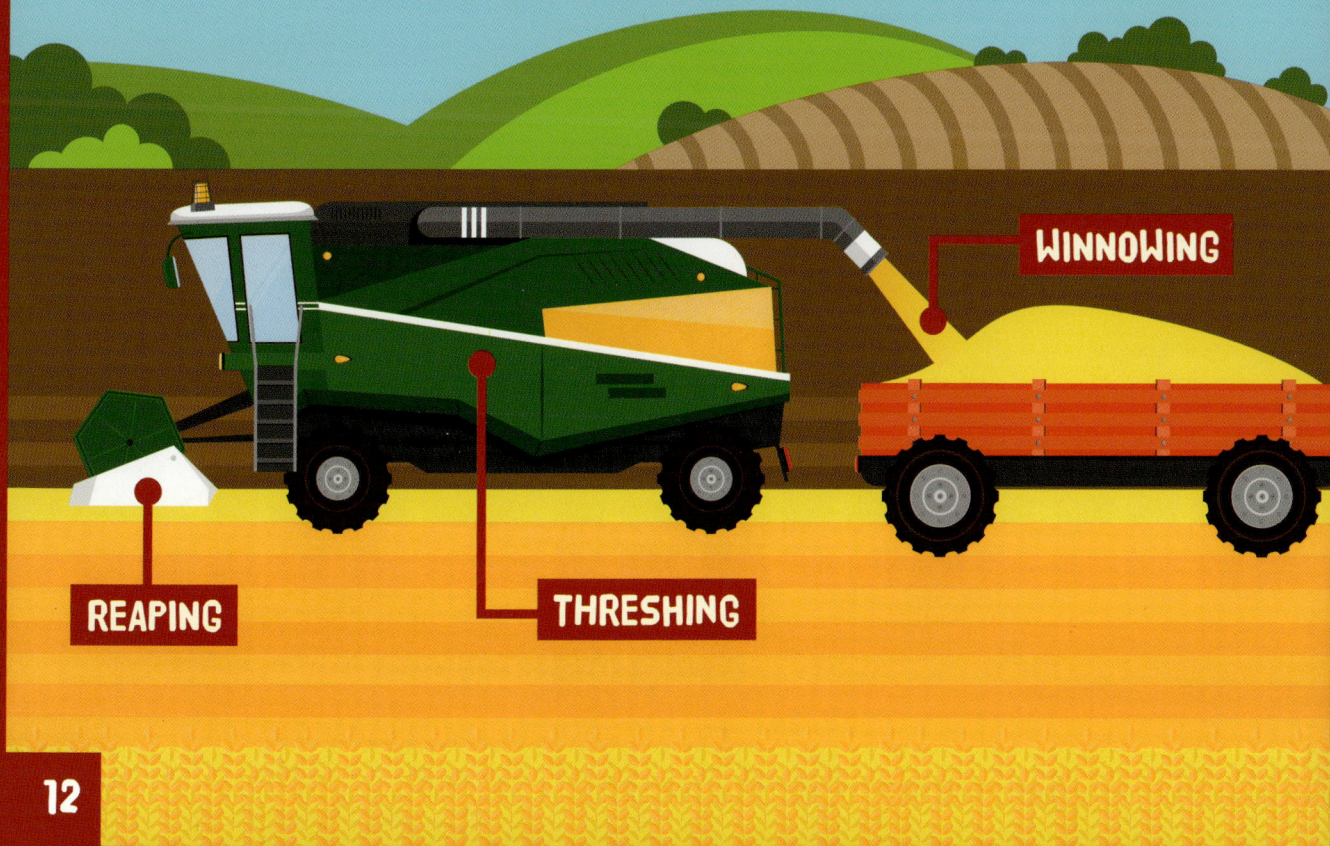

WINNOWING

REAPING

THRESHING

Rice is grown in paddy fields. Rice harvesters are built to be able to work in the wet, muddy fields.

A lot of rice is still harvested by hand.

PARTS OF A COMBINE

Let's look at the parts of a combine.

THRESHER
Inside the machine, large rollers separate grains from the **stalks**.

CAB
Drivers sit in the cab.

REEL
The reel spins and pulls the stalks into the blades.

EXIT PIPE
After being **sieved**, the grains collect in a tank. The stalks fall out the exit pipe and back onto the field.

WHEELS
Big, wide tires help the combine drive over the fields.

BLADES
Sharp blades cut the stems of the plants so the grains can fall in.

The farmer drives the combine up and down the field in rows. Working in rows helps to make sure they don't miss any of the crop.

COMBINES COME IN DIFFERENT SHAPES AND SIZES.

COMBINES OF
THE FUTURE

In the future, a combine might not need a driver at all! The combine could have a computer in it that would tell it what to do.

There are more than seven billion people on Earth. As the world's **population** has grown, we have needed more and more food. Combines have grown bigger!

There could be close to 10 billion people on Earth by 2050!

RECORD BREAKERS

On August 4, 2018, 303 combine harvesters all started up in a field in Winkler, Canada. For five amazing minutes, hundreds of combines worked together!

This broke the world record for the most combines working together!

You could say they combined forces!

On August 15, 2014, a single comine reaped, threshed, and cleaned about 1.7 million pounds (800,000 kg) of wheat!

GET YOUR LLAMA-DIPLOMA

You made it to the end! Well done. Take our little test and you can get your llama-diploma!

Questions

1. Name three crops that can be harvested using a combine.

2. What is another word for cleaning grain?

3. How many stages of the harvest can the combine do at once?

4. What does the reel do?

5. Future combines could go to work without what (or who)?

You made that look easy! Welcome to the Happy Valley Farm family!

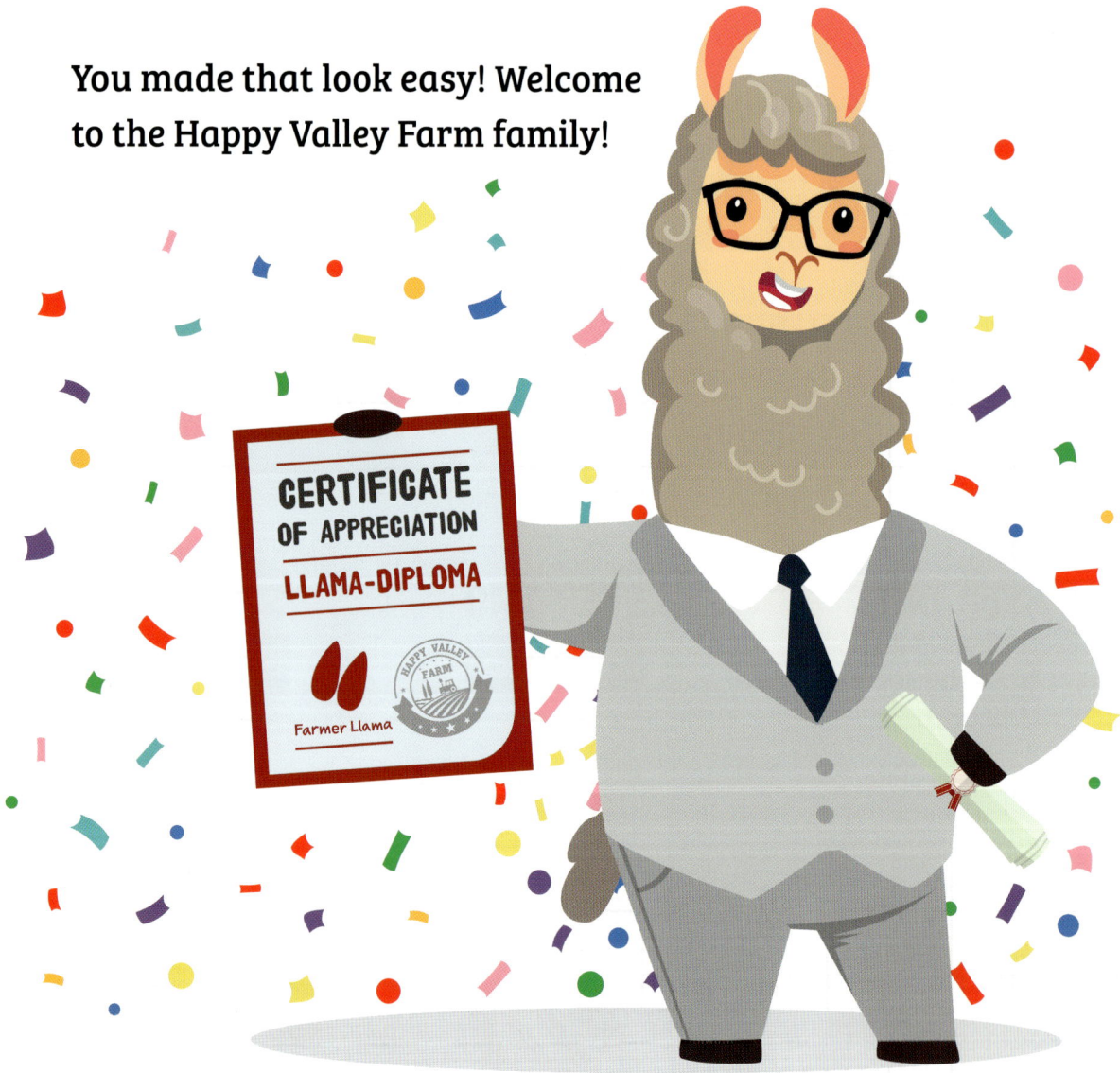

CERTIFICATE
OF APPRECIATION
LLAMA-DIPLOMA

Farmer Llama

HAPPY VALLEY
FARM

Download your llama-diploma!

1. Go to **www.factsurfer.com**

2. Enter "**Combines**" into the search box.

3. Click on the cover of this book to see the available download.

GOING IN CIRCLES

Crop circles are strange shapes that appear in fields. Some people think that they are made by aliens . . .

STEP ONE
Get your combine

STEP TWO
Find a field

STEP THREE
Wait for aliens?

Wow!

But don't forget what a farmer llama can do with an empty field and a bit of time on her hands!

GLOSSARY

BY HAND when something is done by humans without the help of machines

CROPS plants that are grown on a large scale to be eaten or used

FARMHAND a person who works on a farm

POPULATION the number of people in a place

SIEVED passed through a machine in order to separate things

STALKS the main stems of many types of plants

INDEX

COMPUTERS 16

CORN 7

FARM 4, 6, 21

FIELDS 6, 13–15, 18, 22–23

REAPING 8, 12, 19

RICE 7, 13

THRESHING 9, 12, 14, 19

WHEAT 6–7, 19

11/21